PENGUiN PALOOZA

PAIGE TOWLER

FOREWORD BY BERTiE GREGORY
Filmmaker and National Geographic Explorer

NATIONAL GEOGRAPHIC
Washington, D.C.

FOREWORD

Before I began filming wildlife, if someone said the word "penguin" I would think of a small, flightless, tuxedo-wearing bird in Antarctica. It turns out, I had many details wrong! There are 18 penguin species, and each one is full of surprises.

Just five of the world's penguin species live in Antarctica, and only one breeds on ice: the emperor. At four feet (1.2 m) tall, emperor penguins are the world's largest penguin species. They might not have an elephant's memory or an orca's intelligence, but they are incredible athletes. I've watched them huddle together during a minus 40°F (-40°C) storm, keeping comfortable as ice froze on their faces (and mine!). In the water, they are the world's deepest diving birds, disappearing more than 1,800 feet (565 m) down in search of dinner. Their exit back out of the water is no less impressive: They can rocket themselves up to land on ice cliffs towering 10 feet (3 m) above the surface!

Small penguins are no less mighty! Galápagos penguins—the world's second smallest penguins—live where palm trees, volcanoes, and iguanas abound. I couldn't believe my luck when I filmed, for the very first time, a group of them corralling a school of anchovies! The penguins were such efficient hunters that by the time they were done eating they could barely swim with their fish-stuffed stomachs. The penguins bounced off my camera as they struggled to move with full bellies.

Whether they live in the icy wilds of Antarctica or among the lava rocks of the Galápagos Islands, penguins know how to survive in some of the world's most extreme places. And in this book, you'll get to see them doing just that! You'll get to know penguins that live on the ice and penguins that live on sand, big penguins and small penguins, and all sorts of penguins in between. You'll see massive penguin colonies and little bitty, fluffy penguin chicks.

Let the penguin celebration begin!

Bertie Gregory
Wildlife Filmmaker and National Geographic Explorer

You're INVITED, one and all!

It's not a DISCO, nor a BALL ...

But don't delay;

no, don't be tardy:

Come and join the

PENGUiN PARTY!

MEET THE PENGUIN

Say hello to this unusual animal! A penguin is a type of bird. Unlike most birds, penguins cannot fly. That is partly because they have flippers instead of wings. Along with their webbed feet, penguins use these flippers to swim in the ocean.

Grab your gloves and

snag your slippers

(unless you sport WEBBED FEET

and FLIPPERS).

11

It's time to **DRESS UP** to **impress,**

so wear a **CREST**

like **LEMON ZEST.**

PENGUIN PALOOZA

There are 18 different kinds, or species, of penguins around the world. Penguins come in many different sizes. They can also look different from each other. Some kinds of penguins, known as crested penguins, have bright yellow feathers, or crests, above each eye. These crests help them recognize each other and attract mates.

Align your TAIL

and shine your BEAK.

You'll be so **very, very** CHIC!

🐧 PERFECTLY PENGUIN

Like other birds, penguins do not have teeth. Instead, they have beaks. Penguins use their beaks to groom their feathers, catch food, and communicate with each other. Penguins also have big heads, thick bodies, and stubby tails. They sometimes use their tails to help stay balanced.

CHINSTRAP PENGUINS are NAMED for the THIN BLACK LINE under their head, which Looks Like a HELMET'S CHIN STRAP.

CURIOUS COLORS

Penguins of all kinds are famous for their black-and-white coloring. Though this coloring may make them stand out on land, it helps them hide in the water. When seen from above, the black on a penguin's back helps it blend into the dark water below. And to any animal looking up, a penguin's white belly helps it disappear in sunlit waters. This is called countershading. Some other animals—like sharks and dolphins—have countershading, too.

Now **don** your **best**

in **BLACK** and **WHITE,**

you're ready for a

TRUE DELIGHT:

Celebrations are at hand

across a **FROSTY, FROZEN LAND.**

Come see a **scene** that will **entice** ...

Behold: a **SOIREE** on the **ICE!**

WELCOME TO THE ANTARCTIC

Many penguins live in and around the Antarctic. This is the farthest-south region on Earth. It is very cold and often covered in snow and ice. However, it is also a desert! That is because a desert is any place that gets very little rain or snow. Compared to other places on Earth, not many animals live in the Antarctic.

On LAND, a GROUP of PENGUINS is called a WADDLE.

Penguins **SLIP**

and **FLIP** and **SLIDE** ...

...and **ZOOM** and **ZIP**

GETTiNG AROUND

Have you ever wondered why penguins waddle? It might look silly, but waddling helps penguins save energy. And when they need to move fast, penguins slide on their bellies. This is called tobogganing, after the long, flat sled that you might use to slide downhill on a snowy day—just like a penguin!

and **PLOP** and **GLIDE.**

23

NICE AND TOASTY

Because the Antarctic is so cold, it can be hard to live there. But penguins have special ways to keep warm. Penguins have thick, short feathers that trap heat near their bodies. They also have a layer of fat called blubber. This blubber stores heat, too. And if it still gets too cold, penguins can form a huddle! In a huddle, penguins stand close together in a large group to share warmth and keep out the cold.

Or else, if you **prefer** to CUDDLE,

you can **join** a

PENGUIN HUDDLE.

25

Or even **PARTY** on the **sand**;

A little penguin gala's **GRAND!**

The LITTLE PENGUIN is also known as the FAIRY PENGUIN.

🐧 LIFE'S A BEACH

Not all penguins live in the cold! One type of penguin, called the little penguin, lives in Australia and New Zealand. It gets its name from its small size: The little penguin weighs only about as much as two loaves of bread. It is also known for its blue-gray feathers.

And when the **sunshine**

makes things **HOTTER,**

take the **party** **UNDERWATER!**

SUPER SWIMMERS

Penguins are very good at swimming. They use their paddlelike flippers to zip through the water. Penguins can swim very fast, thanks to their special oval shape. This shape lets water move around them easily so they can speed through the ocean. A penguin also uses its feet and tail to help steer while underwater.

Penguins **DIP** and **DIVE** and **DASH**

and **SWIM** and **SWISH**
and **SPLISH** and **SPLASH** ...

In the WATER, a GROUP of PENGUINS is called a RAFT.

FISH DINNER

ALL penguins eat seafood. This includes fish, squids, and tiny shrimplike sea creatures called krill. Because they don't have teeth, penguins swallow their food whole. Backward-facing spines covering their tongues and the inside of their mouths help prevent prey from escaping.

... and **slip beneath** the

To follow 'round a **FISH BUFFET.**

But if you ever **need a** REST,

you can seek out a PENGUIN NEST

in **sorted stones** all **piled neat** ...

NEAT NESTS

ALL penguins lay eggs. Some penguins build nests for their eggs. Some, like the chinstrap penguin, use stones to build these nests. Chinstrap penguins make piles of stones. These rock piles help protect the penguins' eggs until they are ready to hatch.

... or even **between** PENGUIN FEET!

ON THEiR TOES

Other penguins make nests in different ways—and some don't use nests at all. Emperor penguins use their feet! After a female emperor penguin lays an egg, the male balances the egg on his feet. He keeps it warm and safe with his body until the egg is ready to hatch.

And **cozy up** to PENGUIN FLUFF

(at least until you've had enough)!

GROWING UP PENGUIN

When penguin chicks are old enough, they begin to meet each other and make new friends. These chicks explore—but not too far!—and play. They also learn how to swim. Sometimes, young penguins hang out in a large group known as a crèche.

CUTE AND FUZZY

Like all baby birds, baby penguins are called chicks. When penguin chicks hatch, they have soft, fuzzy feathers called down. This down keeps them warm.

Then **hurry out** to **PLAY** some more:

Be **brave**, make **friends**,

and **GO EXPLORE!**

So **don't be late,** try not to DAWDLE.

RUN or SKIP (or even WADDLE).

It's time for

FEATHERED, FESTIVE FUN:

PENGUIN PALOOZA

HAS BEGUN!

PENGUIN GALLERY

There are 18 different species of penguins. Here are the species pictured in this book—and where on Earth they live!

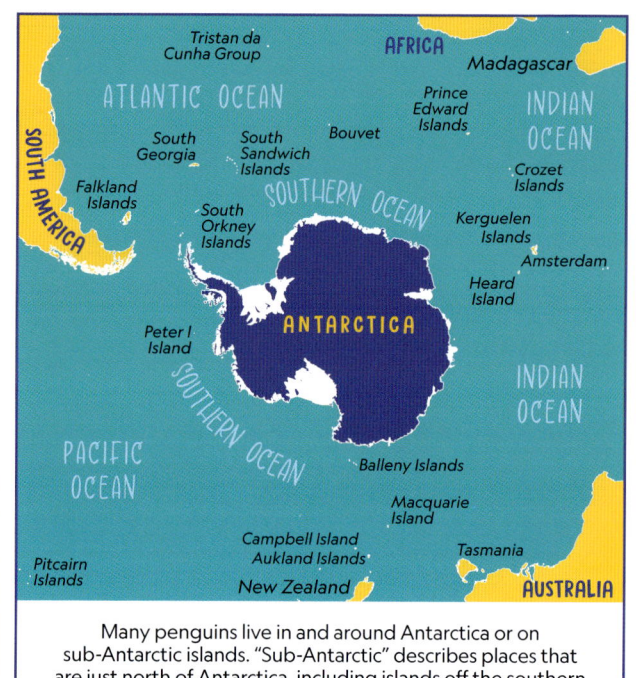

Many penguins live in and around Antarctica or on sub-Antarctic islands. "Sub-Antarctic" describes places that are just north of Antarctica, including islands off the southern coasts of South America, Africa, and Australia. This map shows the Antarctic and sub-Antarctic regions.

Adélie penguin
Antarctica and Antarctic islands

Adélie penguin

Chinstrap penguin
Antarctica and Antarctic islands; islands off the coasts of Argentina and Chile

Emperor penguin
Antarctica

Emperor penguin

Galápagos penguin
Galápagos Islands (Ecuador)

Gentoo penguin
Antarctica and Antarctic islands; islands off the coasts of Argentina and Chile; sub-Antarctic islands off the coasts of Australia and South Africa

Gentoo penguin

Humboldt penguin
Coasts of Chile and Peru

King penguin
Antarctic islands; coasts of Argentina and Chile; islands off the coast of Argentina

Little penguin
Coasts of Australia and New Zealand

Northern rockhopper penguin
Antarctic islands; remote islands in the South Atlantic Ocean

Royal penguin
Sub-Antarctic islands off the coast of Australia

Southern rockhopper penguin
Coasts of Argentina and Chile; Antarctic islands; islands off the coast of Argentina; sub-Antarctic islands off the coast of Australia

GLOSSARY

Antarctic: the most southern region of Earth

blubber: a layer of fat that keeps ocean animals warm

chick: a baby bird

countershading: coloring on an animal that helps it hide in both shadow and light

crèche: a group of young penguins

crest: a raised body part on an animal's head

desert: a region that gets very little rain or snow

flipper: a paddle-shaped body part that helps an animal swim

krill: a small, shrimplike ocean animal

species: a scientific grouping of animals with common characteristics

CREDITS

For Kat: a fantastic editor and friend. —PT

Since 1888, the National Geographic Society has funded more than 14,000 research, conservation, education, and storytelling projects around the world. National Geographic Partners distributes a portion of the funds it receives from your purchase to National Geographic Society to support programs including the conservation of animals and their habitats. To learn more, visit natgeo.com/info.

For more information, visit nationalgeographic.com, call 1-877-873-6846, or write to the following address:

National Geographic Partners, LLC
1145 17th Street NW
Washington, DC 20036-4688 U.S.A.

More for kids from National Geographic: natgeokids.com

National Geographic Kids magazine inspires children to explore their world with fun yet educational articles on animals, science, nature, and more. Using fresh storytelling and amazing photography, *Nat Geo Kids* shows kids ages 6 to 14 the fascinating truth about the world—and why they should care. natgeo.com/subscribe

For rights or permissions inquiries, please contact National Geographic Books Subsidiary Rights: bookrights@natgeo.com

Designed by Amanda Larsen

Library of Congress Cataloging-in-Publication Data

Names: Towler, Paige, author.
Title: Penguin palooza / by Paige Towler.
Description: Washington, D.C. : National Geographic Kids, [2025] I Audience: Ages 5-8 I Audience: Grades K-1
Identifiers: LCCN 2024012939 I ISBN 9781426377150 (hardback) I ISBN 9781426377761 (library binding)
Subjects: LCSH: Penguins--Juvenile literature.
Classification: LCC QL696.S473 T678 2025 I DDC 598.47--dc23/eng/20240427
LC record available at https://lccn.loc.gov/2024012939

The publisher would like to thank Katherine Kling for fact-checking this book, Dr. Michelle LaRue for her expert review, and the book team: Kathryn Williams, editor; Lori Epstein, photo manager; Colin Wheeler, photo editor; Mike McNey, map production; Alix Inchausti, senior production editor; Yogi Carroll, production manager; and Lauren Sciortino and David Marvin, associate designers.

Printed in China
25/HHC/1